COMMAND OVER TECHNOLOGY

DIGITAL TRANSFORMATION IS A PROCESS
THAT HUMAN, NOT TECHNOLOGY, DRIVE.

BOSCO EKKA

Contents

Contents

Foreword

Forward

Students are becoming more interested in jobs in artificial intelligence and are incorporating the subject into their curriculum. Data scientists come from various educational backgrounds; even if you have previous experience with coding and program development, you must understand all aspects of data science to become a data scientist. Numerous for-profit and not-for-profit institutions across Australia offer diploma, graduate, postgraduate, and certificate programs in and on artificial intelligence. CryptoInvestingInsider.com walks you through the process of purchasing a high-quality 4K portable monitor. Never invest in bitcoin with money you can not afford to lose.

Saudi Arabia's Interior Ministry wants to establish a biometric system based on iris recognition. Employment is projected to create approximately 2 million jobs by 2030. Artificial intelligence contributes significantly to the development of new job possibilities in the technology sector. Businesses are expected to place a premium on digital transformation soon, but many will fall short of realizing the technology's potential. According to Andrew Dolan, digital transformation is a human-driven process, not a technical one.

While technology is all about doing more with fewer resources, he asserts that it is most successful when combined with human capabilities. Dolan Invests in training to maximize effect in high-value sectors. PC Care's extensive service portfolio includes assistance with Linksys routers, operating systems and software programs, email and browser support, and peripheral device installation assistance. The goal of PC Care is to guarantee your comfort and the effective functioning of your computer by providing individualized online computer support services that address your laptop's overall health. The Sonos ecosystem is a great way to enjoy music in crystal clear surround sound across your home.

You may link Sonos to an existing Crestron panel through a Sire gateway. A professionally developed iOS application enables you to promote your brand or business by broadening your market reach, protecting consumer information, engaging customers, and increasing productivity.

Preface

Preface

Data science is a collection of techniques and studies that may extract desired information from a set of unprocessed or raw data. It is not a single topic but a broad area with many components and facets that must be understood to advance in the data science profession. Data scientists come from a diverse range of educational backgrounds; even if you have prior expertise in coding and program development, you must grasp all facets of data science to become a data scientist. AI is a computer-based model of human intelligence. Computer science is concerned with developing intelligent robots and technologies capable of responding and reacting in a manner consistent with humans in real-world circumstances.

Students are increasingly interested in careers in artificial intelligence and are studying the topic as part of their education. Students may benefit from data science courses in Australia by increasing their knowledge of artificial intelligence. Numerous for-profit and not-for-profit schools across Australia provide diploma, graduate, postgraduate, and certificate level courses in and on artificial intelligence. These classes include a broad range of subjects, including physics, mathematics, technology, data science, computer science, and neurology. Admissions are straightforward.

Each institution has its own admissions requirements and may require you to take an entrance exam to show your proficiency. You may contact us if you're searching for an artificial intelligence course in Australia. If you're searching for a cable gland that is resistant to water, we recommend placing one near a source of water. If electrical grounding or earthing is required, ensure that you choose high-quality equipment that has this function. If the cable is armoured, consider the inner bedding, the material, and the armour's short-circuit ratings.

Additionally, the diameter and depth of the gland's thread should be checked. If your project requires armoured cables, we recommend that you examine several critical aspects before investing in cable glands. You must establish the kind of armour and the resources needed to build it. You should verify the cable's current rating, lead coating, and inner bedding sizes. If you install these units in a high-risk location, ensure that you adhere to local safety requirements.

Cable Glands are installed at the cable's entrance locations to maintain the cable's ingress protection grade. They are often constructed of metallic or non-metallic materials and are resistant to corrosion. Shenzhen-based Flex Glory Cable Accessories Co., Ltd. manufactures high-quality cable glands for NPT and PG cables. According to the firm, its devices provide superior EMC performance and Earth continuity. A cable gland is a device that is attached to the end of a properly installed electrical cable.

Cable strain reliefs, cord grips, cable fittings, and cable enclosures are also referred to as these. The primary function of these devices is to ensure the safe movement of wires and cables inside an enclosure. They are utilized in various applications, including telecom cables, data transmission, measurement, control, and power. The number of banks that have banned cryptocurrency transactions using their credit cards continues to rise. Chase, Bank of America, and Citigroup have all jumped on board with this new trend of crypto purchasing limitations.

Banks do not want individuals to spend large sums of money that will be difficult to recover in the event of a significant cryptocurrency collapse. It protects the customer by lowering the risk of financial difficulties due to utilizing credit to purchase something that leaves them cash and credit poor. CryptoInvestingInsider.com demonstrates how to invest in a high-quality 4K portable monitor. When purchasing a 4k portable monitor, size, weight, and display quality are critical factors. If you need a larger screen to watch Netflix, you may not require a larger screen.

By investing in bitcoin, you should never risk money you can not afford to lose. Adjusting the panel's orientation is a critical feature to look for in a portable 4K monitor. It will be easier to see the whole picture if you rotate the display from portrait to landscape mode. The portrait orientation makes huge blocks of code easier to scan. If your desktop monitor lacks in this region, an error message may appear on the screen.

Intelligence Artificial Artificial intelligence is the idea of developing technology that operates, acts, and responds in a manner comparable to that of a human person in real-world situations. By 2030, the employment industry is expected to generate over 2 million jobs. Numerous businesses realize that artificial intelligence makes a significant contribution to creating new employment opportunities in the technology industry. After completing your artificial intelligence course, you may pursue employment in these fields. The Saudi Arabian Interior Ministry wants to implement an iris recognition biometric system.

The device utilizes mathematical recognition methods to scan a person's iris before granting entry to the facility. It is intended to deter unauthorized individuals from obtaining access to a protected location. The government is now investigating the circumstances behind the installation of these security gadgets. Face recognition technology may contribute to the abolition of the need for traditional fingerprint identification. These gadgets will scan and validate your face in a matter of seconds.

Once installed, these face recognition technologies are incapable of being abused. If you are an employer, these methods will be very beneficial. FirstCity is the market leader in this security system category. Businesses are anticipated to emphasize digital transformation soon, yet many fail to capitalize on the technology's promise. Digital transformation is a human-driven process, not a technological one, and it is very difficult to change a company with faulty organizational procedures.

The Covid-19 pandemic will be "the infiltration of data-enabled services into an ever-increasing variety of areas of existence." In today's competitive market, digital transformation is essential to a business's ongoing relevance and profitability. It necessitates the development of new tools and apps, the storage of data, the recording of information, and the adoption of many new techniques. While you can not delegate substantial decision-making authority to your team, engaging them in the process may enhance results. The purpose of digital transformation services is to make life easier and more enjoyable for individuals. Consult with your employees if you're considering using a new platform for online collaboration but aren't sure which Zoom, Teams, or Slack to use.

By investing in training, you may accelerate this process via online courses and hands-on learning. The CEO or top leader of a company is the single most significant element affecting the transformation's efficacy. While technology is all about doing more with fewer resources, it is most effective when coupled with human skills. The significant conclusion is that when leaders evaluate new technology, they also need to consider the people who enable it to be helpful. While the rest of an organization's operations may be duplicated, talent can not. Invest in the finest talent to maximize impact in the areas that create the most value.

AngularJS was created more than a decade ago and has seen many modifications since. Misko Hevery, a programmer who took on a side job, created the framework. The framework's initial version was released in 2009, soon after the release of AngularJS. After quickly gaining popularity alongside other frameworks, it attracted Google's attention. The Prince2 Agile Foundation is a valid certification and qualification program for anybody working in an agile project environment.

To get the certificate, applicants must properly answer test questions demonstrating their knowledge of each topic. Individuals with no prior understanding of PRINCE2 cannot sit for the exam since the technique is taught throughout the course. To learn more, please visit gotechblog.com. Candidates must develop and follow a study plan covering the bulk of the curriculum's subjects. Keep informed and connected to a PRINCE2 network at all times.

Establish a goal and adhere to it. It may provide candidates with more time to practice the subjects included in their study plan. Contact our advisory team if you're interested in learning more about the advantages of project management for your company. PC Care's comprehensive service portfolio includes Linksys router help, operating system and software program assistance, email and browser help, and assistance with peripheral device installation. PC Care's mission is to ensure your comfort and the efficient operation of your computer by offering personalized online computer support services that address the entire health of your laptop.

While Crystal Report Writer enables you to create aesthetically attractive reports, you may not find a more efficient delivery method unless you have a flexible software program. While locating instruments that fulfill these criteria may take some time, the effort is worth it. You may generate and suppress document dissemination according to your needs using a customizable document bursting tool. The Miami web design firm has you covered from simple site design to search engine optimization or SEO. We provide a thorough study of your SEO keyword rankings, a report on your link-building profile, and information regarding indexed sites.

Our customer-focused team will increase the traffic and income of your website. The competition for digital technology advances is on; recent developments in digital technology, particularly smartphones, have altered the landscape. Smartphones are the most effective tool for boosting sales in your company. The following are the main reasons for creating business apps on iOS. Apple tightly controls the whole ecosystem, from hardware to firmware to programming.

A professionally designed iOS application allows you to promote your brand or company, expanding your market reach, safeguarding your customers' information, engaging your clients, and boosting productivity via services that connect you with your targeted consumers. The Sonos ecosystem is an excellent way to enjoy music across your house in crystal clear surround sound. With the Sonos app and Sonnex system, Crestron's audio transmission improves this experience even further. Through a Sire gateway, you may connect Sonos to an existing Crestron panel. The versatility and increasing popularity of the Sonos system, along with Crestron's cutting-edge home automation technology, work exceptionally well to bring your home into the modern-day.

CHAPTER ONE

The Importance of Data Science in Careers

What does the term "Data Science" mean?

In terms of utility, data has exceeded all other areas of our existence. Every day, you transmit and receive millions of bytes of data without even realizing it. Whether you're working, playing, or just watching a movie, data transfer underpins everything. Given the importance of data in our lives, it's not surprising that data science has emerged as one of the fastest-growing career options for skilled people.

To be clear, it is not a single subject; rather, it is a collection of methods and studies that may be used on a set of unprocessed or raw data to extract the desired information. It may involve mathematical modeling, statistical techniques, and even basic business acumen that are utilized to provide data to unearth hidden information or trends that assist a firm in determining its next move.

It is very adaptable since it uses past data to determine the optimal course of action in the present and even creates future models.

What Motivates You to Pursue a Data Science Career?

The internet has eased almost every area of our lives. That is why every sector in the modern era uses data science to propel their businesses forward in the digital era. This is important to understand since many companies and organizations use the internet to do business. Consequently, these businesses now have access to a huge quantity of unprocessed data they want to profit from. As a result, they need a data scientist to handle the data properly and extract critical information from it.

Additionally, this is advantageous since data scientists come from a variety of educational backgrounds. Even if you have previous experience with coding and program development, you must understand all aspects of data science to become a data scientist. That is why, in recent years, the data scientist job has become the most sought-after.

The Data Science Components

It is not a single subject but rather a wide field with many components and aspects that must be grasped to progress in a career in data science. Several of data science's most significant subfields include the following:

Extensive Data

The phrase "Big Data" refers to any unstructured or raw data generated or collected anytime a digital medium is utilized. When properly analyzed, this data is important for various organizations and commercial activities since it may provide information about consumer patterns and predict future actions.

Intelligence Artificial Intelligence

Over the past several years, machine learning has risen in popularity among students interested in data science. In general, it involves applying mathematical and statistical concepts to raw data to extract the desired information.

This data may be used to assist a business in creating its marketing plan.

Entrepreneurial spirit

The data scientists analyze the data, but not the people who make business decisions. To aid in their understanding, collected data must be transformed into a more understandable visual representation.

The popularity of data has increased the number of institutions that provide it. If you're interested in learning more about data science courses in Australia or data science courses in Melbourne, you can do so by visiting their websites.

CHAPTER TWO

Fundamentals of Artificial Intelligence—Careers, Admissions, and Requirements

What does the term "Artificial Intelligence" mean?

Artificial Intelligence (AI) is a computer-based model of human intelligence. Computer science focuses on creating intelligent robots and technologies capable of behaving and reacting similarly to humans in real-world situations. Artificial Intelligence may use several techniques to achieve a mirror image of human cognition. Reasoning, self-correction, and learning are only a few of the processes that constitute human intelligence. That is why a large number of people enroll in an AI course in Melbourne.

As part of its learning process, AI has to gather knowledge and learn how to utilize it. AI takes advantage of these concepts to get the most precise outcomes possible. AI applications include speech recognition, planning, problem-solving, learning, expert systems, and machine vision. Certain computers are constructed with the help of artificial intelligence to perform certain jobs and activities.

Introduction to Artificial Intelligence for Students

It is a subfield of computer science focused on developing intelligent robots capable of moving and speaking like humans. Artificial Intelligence has developed into a key component of the technology industry over the years.

Students are increasingly considering jobs in AI and studying the subject as part of their schooling. As a consequence of all of this, demand for AI courses has increased significantly. Before continuing, it is essential to realize that research into artificial intelligence is very technical and specialized. Students often pursue courses to gain knowledge regarding problems relating to artificial intelligence, such as the programming of features such as learning, mobility, and problem-solving.

Courses in Artificial Intelligence (AI)

When it comes to finding artificial intelligence courses, it's important to note that they're available at various levels. However, since AI was discovered relatively recently, its applications are still in their infancy. Its youth shows how quickly its importance has risen over the past several years.

Additionally, several institutions have created their own AI courses in response to the development and popularity of AI. Internationally recognized universities have included studies of artificial intelligence in their curriculum. And this is not restricted to colleges of science or technology. Numerous educational institutions are experimenting with artificial intelligence education as well.

These AI courses include a wide variety of disciplines, including physics, mathematics, technology, data science, computer science, and neurology. Numerous institutions across Australia, both for-profit and not-for-profit, provide diploma, graduate, postgraduate, and certificate level courses in and on artificial intelligence. Obtaining a Master of Science (M.Sc), a Bachelor of Technology (B.Tech), or a Master of Technology (M.Tech) with a focus on artificial intelligence is now simpler than ever (AI). Interested students may also enroll in remote learning programs or short-term artificial intelligence courses offered by certain institutions in Australia.

Admissions are simple. It starts with selecting a suitable institution and continues with the completion of the registration form and process. Each school has its admission criteria and may even ask you to take an entrance test to demonstrate your expertise. It guarantees that the admissions process is fair for all students interested in pursuing subsequent AI courses.

If you're looking for an artificial intelligence course in Australia, you may contact us. Courses in artificial intelligence provide the ideal learning environment. Learn about the curriculum and admission criteria for Melbourne's AI program and prepare to join the future of data science.

CHAPTER THREE

Consider the Following When Purchasing Cable Glands

Nowadays, cable glands come in a range of designs and sizes. As a consequence, you must study and choose the best possible kind. However, before making a choice, we suggest that you consider many important factors. We'll explore some of the factors that may help you make this choice in this article. Continue reading for more information.

To begin, choose the gland that is suitable for the cable type. For instance, if you have an SWA armored cable, you will need a gland comparable to the BW brass cable gland.

Additional Points to Consider

Another important aspect to consider is the material used to build the cable. Additionally, you must ascertain the cable's braided or screened construction. Apart from that, you must determine the cross-sectional dimensions and structure of the unit.

Additionally, you must consider the gland's colour needs, especially if you place a premium on aesthetics. Determine if there are any limitations on the installation space, the surroundings, or electromagnetic interference. Additionally, you must consider the location of the gland and cable.

We suggest installing a water-resistant gland if your system is located near a source of water. The IP68 rating implies that the gadget is waterproof and sealed against dust. Similarly, IP69K-certified gadgets are waterproof and will continue to operate properly if immersed in water.

You've arrived at the right spot if you're looking for a gland that will offer mechanical protection.

If the unit must be used in a dangerous area, such as one containing explosives, we suggest using a unit with a safe circuit, such as the HSK-K-EXE-Active.

If you need electrical grounding or earthing, be certain you choose high-quality equipment that has this feature. Determine if any problems will arise as a result of the reactivity of certain metals.

If the cable is armoured, consider the inner bedding, the material, and the short-circuit ratings of the armour. Additionally, consider the gland's housing or mating substance. This is essential if you want to ensure that the components function correctly in conjunction with one another.

Additionally, the diameter and depth of the gland's thread should be examined.

At times, stopper plugs in the gland are required to close off unnecessary cable entries. As a consequence, you should ascertain whether or not you need this kind of stopper plug.

To summarise, if you're looking for cable glands, we suggest you consider the factors mentioned in this article. Apart from that, you may want to seek the advice of a technical expert. It will aid you in making the best cable gland selection.

Are you looking for the best deal on a PG cable gland or a waterproof cable gland? If this is the case, we strongly advise you to contact Shenzhen FlexGlory Cable Accessories Co., Ltd.

CHAPTER FOUR

Suggestions for Selecting the Appropriate Size of Cable Glands for your Project

If you're considering purchasing cable glands, we suggest taking your time to avoid future errors. Consequently, many factors, including the size and kind of cable glands, must be considered. After all, the last thing you want is to end up with a cable or gland that is the wrong size. You need an item that fits properly. In this article, we're going to provide some recommendations to help you choose the right size cable gland for your project. Continue reading for more information.

While this choice may be made at any time, the ideal time is during the planning phase. It will help you to avoid problems throughout the gland size determination process.

Two important factors must be addressed while choosing the cable fitting size. To begin, consult with your cable technician to identify the proper cable for the task. The next step is to calculate the diameter of the cable and the cable glands.

If your project necessitates the use of armoured cables, we suggest that you consider many important factors before investing in cable glands.

To begin, you must ascertain the kind of armour and the materials used in its construction. Following that, check the current rating, lead covering, and inner bedding diameters of the cable.

Additionally, you should decide whether you require a cable gland with a protective coating for corrosion protection. In general, if you are needed to work in particular circumstances, you must consider these factors. Apart from that, you should learn about the materials used to construct electrical enclosures. If you do not use appropriate metals, there may be serious repercussions.

Another important element to consider is the cable's entry hole, especially its type and size concerning the electrical equipment it will be attached. Additionally, you should check the entry thread seal, shrouds, and earth tags, to name a few. Occasionally, you will also require stopper plugs or reducers.

Researching installing the cables will help you avoid the problem of choosing the wrong cable gland size. Apart from that, doing your research may speed up the installation procedure. If installing these units in a high-risk area, verify that you comply with the local authorities' safety regulations.

To conclude, if you wish to select the right cable gland size for your project, we suggest that you follow the recommendations given in this article. It will help you to avoid common mistakes when making this decision.

Hopefully, these ideas will help you to stay safe and make an informed decision.

You may obtain your selected Metric cable gland from Shenzhen FlexGlory Cable Accessories Co., Ltd. Additionally, and they offer the best waterproof cable glands.

An Overview of the Cable Gland

Cable glands are equipment used with wires and cables in automation systems such as telecommunications, data, power, and lighting. Their primary function is to serve as a terminating and sealing unit, ensuring the safety of electrical enclosures and equipment. We're going to give you a quick overview of these units in this article. Continue reading for more information.

Protection of the Environment These devices protect the environment by enclosing the outside cable sheath, thus preventing dust and moisture from entering the instrument or electrical enclosure.

Earth Continuity: This term refers to the use of shielded cables in combination with earth continuity. This kind of cable gland is made of metal. As a result, these components are tested to ensure they can withstand the maximum short-circuit fault current. They are used to provide a greater gripping force to withstand the resistance.

Additional Sealing: It serves as an additional seal when a high degree of ingress protection is required.

Cable Glands: Cable glands are used at the cable's entry points. The objective is to maintain the Ingress protection rating through the use of the right accessories.

They are often made of metallic or non-metallic materials. The majority of the units are corrosion-resistant. As a result, each item is subjected to corrosion resistance testing.

If you want to use these units in potentially explosive settings, you may wish to use glands that are specifically developed for the cable type you intend to utilize. The following are some of the main characteristics of high-quality cable glands.

(1) Self-contained inner sealing

A quality unit utilizes a unique inner sealing concept in comparison to other kinds of cable glands. The internal sealing ring is separated from the armour clamping components, reducing the risk of over-tightening.

Compression seals utilized conventionally do not provide direct control over their application. However, this kind of inner sealing method utilizes a displacement seal that is independently regulated during installation.

(2) Termination of Secure Armour

This termination method may assist in ensuring that the armour is permanently crimped. And this results in a low-impedance connection that is self-tightening-free. The clamping ring makes installation simple. Apart from that, you'll benefit from outstanding EMC performance and Earth continuity.

(4) Seal on the outside

The distinctive outer seal of these glands ensures a good seal around the cable. The exterior seal is sufficiently robust to resist considerable pressure in some situations.

(5) Flood-proofing

The deluge seal may help prevent rusting. It is feasible because moisture does not accumulate on the ring of the cable gland. Apart from being resistant to mechanical stress, the O ring is also resistant to damaging UV radiation.

In closing, this short introduction to cable glands should have helped you better understand them.

CHAPTER SIX

Why Is It Required To Use Cable Glands When Connecting Electrical Equipment?

A cable gland is a device connected to the end of a good electrical cable that connects to a device in its simplest form. These are also known as cable strain reliefs, cord grips, cable fittings, and cable enclosures. In this post, we'll examine why you should utilize these units for your devices, including switchgear and other electrical equipment. Continue reading for more information.

To begin with, these glands serve as a means of terminating cables in potentially dangerous locations. For instance, this is a crucial region in which sufficient ingress protection is required. These cable glands are essential for strain relief and attaching the cable to the cable armour and aluminum sheath.

Cable glands are usually constructed of metallic, non-metallic, or a mix of the two elements. They are used in several sectors where automation systems and electrical instruments need a mix of wires and cables.

These are available in a variety of cable kinds and lengths. The main purpose of these devices is to guarantee the safe flow of wires and cables inside an enclosure. They are, nevertheless, needed to safeguard electrical components from fires.

The Goal

These glands are a crucial component of an effective electrical system design. The main purpose of these devices is to strengthen cable attachments. They withstand dust and twisting.

The good news is that these glands can be used in a wide range of applications, including telecom cables, data transmission, measurement, control, and power. As a sealant and termination unit, they may aid in the appropriate maintenance of enclosures.

Why Is It Required to Obtain Cable Glands?

Cable glands provide several protective purposes for equipment. They may be used to alleviate tension, insulate, bind the soil and ground. Additionally, they may help seal cables that pass through a variety of gland plates and bulkheads. Additionally, they help protect the connection from becoming clogged with dust, debris, and fluids.

How do they work?

They are mainly used as a sealing device to safeguard the enclosures of the equipment. Additionally, they contribute to sealing the intake site, preventing extraneous particles from harming the cable and system.

Furthermore, environmental pollutants such as water, fluid, and soil may damage the cable joints. Thus, cable glands are used to prevent the cable from being tugged or twisted.

Typical Uses

These devices ensure the continuity of the earth in a cable type called armoured cable. Additionally, if the cable gland is constructed of metal, it increases the cable's resistance to high-current fault currents.

Additionally, it serves as a stabilizing force, ensuring that adequate resistance exists. By creating a seal on the outer sheath, it protects the surroundings. It effectively seals off the sealed area from dust and moisture.

They may aid in sealing sections of the system that need a high degree of protection against intrusion. Additionally, they safeguard the environment near the cable's entrance site.

These are just a few of the reasons why cable glands are a good choice for your systems.

Investing in high-quality cable glands is essential to ensure your equipment operates properly. These devices are available in several configurations, including metallic cable glands, PG cable glands, and others.

CHAPTER SEVEN

Why Have Banks Banned Credit Card Purchases of Cryptocurrencies?

The number of banks prohibiting cryptocurrency purchases with their credit cards grows, with Wells Fargo entering the fight. Several different financial companies, including Chase, Bank of America, and Citigroup, have joined this new crypto purchase restrictions trend.

While debit cards appear to be able to be used to purchase cryptocurrency (check with your bank to confirm their policy), the use of credit cards to purchase cryptocurrency has taken a turn, with these banks leading the way with these purchase bans, which are likely to become the standard soon.

When credit cards were used to buy cryptocurrency, apparently overnight transactions started to be canceled, and people who had never had any difficulty acquiring cryptocurrency with their credit cards discovered they were no longer allowed to do so. The bitcoin market's volatility is to blame. Banks do not want people to spend huge amounts of money that will be impossible to return if a major cryptocurrency crash happens, as it did earlier this year.

Of course, these banks will also lose money if people purchase cryptocurrencies and the market increases, but they seem to have decided that the downside of this risk with their credit cards outweighs the gain. Additionally, this protects the consumer by reducing the likelihood of getting into financial trouble due to using credit to buy something that will leave them cash and credit poor.

The bulk of investors who purchased cryptocurrencies using credit cards did so ostensibly search for fast gains and had no intention of sticking in for the long term. They planned to enter and leave quickly, then pay off their credit cards before the astronomical interest rates taking effect. However, due to the cryptocurrency market's ongoing volatility, many investors who made this transaction ended up losing a substantial amount of money after the market's collapse. They now have to pay interest on the money they lost, which is never a good thing. Naturally, this was bad news for banks, and it led to the current and growing trend of credit card issuers banning cryptocurrency purchases.

The lesson here is that you should never spend all of your available credit on crypto and only utilize a part of your hard assets to purchase crypto. This money should be a long-term investment that will not jeopardize your financial position.

As a result, avoid investing funds in bitcoin that you may need soon, only to find that a downturn has depleted your wallet. According to an adage, "Do not gamble with money you can not afford to lose," and this is the lesson banks want customers to learn as they go into this new financial frontier.

While certain methods for investing in cryptocurrencies are right and others are wrong, purchasing bitcoin with a credit card is feasible if done properly. CryptoInvestingInsider.com shows you how to make bit-coin investments.

Characteristics of a 4K Portable Monitor of Superior Quality

Admissions, Requirements, and Additional Information for Artificial Intelligence as a Career

As demand for AI increases, many colleges and organizations in Toronto, Canada, have started offering AI courses at various levels. The importance of artificial intelligence has grown significantly in recent years. That is why AI courses are still in their infancy.

Numerous renowned universities have expanded their curriculum to include the courses in response to their increasing popularity and demand.

And this is not limited to scientific or technical colleges; other educational institutions are also involved. You will come across the phrase 'Machine Learning' when pursuing a profession in artificial intelligence.

What is the difference between Artificial Intelligence and Machine Learning?

While the terms AI and machine learning are often used interchangeably, they are different ideas that should not be confused. Furthermore, their interchangeable usage often contributes to uncertainty for those trying to understand the true meaning of these terms. Another misconception about these two is that they function similarly.

In reality, artificial intelligence is the concept of developing robots capable of mimicking human behaviour and doing tasks intelligently.

On the other hand, machine learning is an application of artificial intelligence that gives computers full access to information and data for them to self-learn.

Artificial intelligence aims to increase the likelihood of success while minimizing the amount of effort needed for accuracy. Machine Learning accomplishes the reverse, concentrating only on accuracy and disregarding success rates.

Career Opportunities in Artificial Intelligence

After you've established the difference between Machine Learning and Artificial Intelligence, you must evaluate your career options. Additionally, job possibilities are expected to increase to 2.3 million positions by 2030. AI is and will continue to be a driver of new job development in the technology industry.

As such, if you're interested in pursuing a career in Artificial Intelligence, the following courses are available:

1. Machine learning research
2. Video game programmer
3. Data analyst 4. Software developer 5. Robotics programmer
6. Business intelligence developer
7. Data mining analyst
8. Researcher in Science

9. Military Forces

Artificial Intelligence's reach is growing every day, and there are many employment possibilities in this area. If you possess a thorough knowledge of artificial intelligence, you may anticipate working with major corporations such as Google, Amazon, Facebook, Uber, Microsoft, and IBM, among others.

How Can an Artificial Intelligence Certification Help Your Career?

Before participating in a program, you must ascertain your eligibility. To work in the field of artificial intelligence, you must have a solid understanding of mathematics and computers.

If you are new to artificial intelligence, it is recommended that you begin with mathematics and then go on to take machine learning courses. Additionally, you should possess programming skills and a basic understanding of algorithms. If you already have some experience or abilities in programming, you may go straight to the code and algorithms.

Additionally, regardless of what institution or university in Canada you choose for an AI degree, you must be prepared to acquire new skills throughout your career in AI continually. Why should you, given that machines never stop learning?

Are you considering the merits of a career in artificial intelligence? However, there is a reason why you should enroll in a Canadian artificial intelligence school. Determine your eligibility for artificial intelligence courses in Toronto, California, and apply to the best school for a professional start.

CHAPTER TEN

How Will Artificial Intelligence Courses Fare in the future?

As the need for Artificial Intelligence increases, more students enroll in Cape Town schools that offer AI degrees. Under the aegis of artificial intelligence, one may find a plethora of college and course options. The ever-expanding technological sector needs the expertise of specialists.

If you're unfamiliar with the term or are considering a career in this field, you've arrived at the right place. We will educate you about the many avenues available for pursuing a career in artificial intelligence.

Are Artificial Intelligence and Machine Learning synonymous?

While Artificial Intelligence and Machine Learning are often used interchangeably, they are separate terms with distinct meanings. To define AI, one might say that it is the concept of creating technology that functions, behaves, and responds similarly to a human being in real-world circumstances. On the other hand, machine learning is based on the concept of self-learning, in which you provide data to the machine, and it learns on its own. Machine learning is a subset of artificial intelligence in which the machine analyses and learns from its experiences via the use of algorithms. While the main goal of machine learning is to improve accuracy, the rate of success is often neglected. Artificial intelligence makes up for the low success rate by raising the probability of achieving it.

Intelligence Artificial Opportunities for Employment

The application of artificial intelligence is growing at an exponential pace. By 2030, it is anticipated that the employment sector will create over 2 million jobs. Most companies recognize that artificial intelligence contributes significantly to creating new job opportunities in the technology sector. And the best part is that you may apply for a range of professions after your studies are completed.

Numerous employment opportunities exist in artificial intelligence, including the following:

1. Robotics-focused programmer
2. Data mining analyst 3. Video game programmer
4. Engineer with expertise in machine learning.
5. Machine Learning Research
6. Researcher in Science
7. Business intelligence developer 8. Software developer
9. Armed Forces

Engineer in Artificial Intelligence

You may seek work in these areas after finishing your artificial intelligence training. Salary ranges may vary by position or industry, but they are worth exploring.

The Future of Artificial Intelligence

It's impossible to forecast what will be significant in the coming years. Each day, new technical breakthroughs occur. As a result, there is no definitive answer to whether or not entering the field of artificial intelligence is a wise career choice. However, there are many reasons why one should immediately pursue it. The industry is eager for more educated and experienced professionals in artificial intelligence, which is pushing up demand. Currently, the industry is experiencing a skills shortage. Thus, getting certification in an AI course and starting your studies is a great career move.

In general, admissions to colleges and institutes offering artificial intelligence courses in South Africa are simple. It starts with an eligibility check and completes an application form, and administering an entrance test (if necessary). Keep in mind that you may need to develop communication and business talents before entering the job market in addition to arithmetic and computer skills.

If you're interested in pursuing an Artificial Intelligence career in South Africa, you may begin by enrolling in a Cape Town AI course. Examine the course options that interest you and develop the skills required for business success.

CHAPTER ELEVEN

Launching of Face Recognition System

According to sources, the Saudi Ministry of Interior wants to deploy an iris recognition biometric system. It covers plans for major sites such as airports, seaports, and land. The new systems will make use of cutting-edge technology for passenger identification. In other words, this cutting-edge technology is designed to prevent unauthorized people from gaining entry to a secure area.

The National Information Centre of the government will be responsible for deploying the iris recognition technology. Essentially, this technology scans a person's iris before allowing access to the facility using mathematical recognition techniques.

The National Information Centre will enter into a deal with the Saudi Arabian government to import and deploy the technology. After installation, this system will function at a high degree of efficiency consistent with international standards. At the time, potential bidders had been invited to submit their bids. Besides that, the terms and conditions will be addressed. Apart from that, the government is investigating the conditions and circumstances surrounding installing these security devices.

How is the System of Facial Recognition Implemented?

In essence, the Iris recognition system is a more advanced version of the most commonly used biometric identification technique based on fingerprints. A medical study indicates that each person has distinct and distinctive ocular features. And the good news is that no two people's iris features are similar. Indeed, everyone's eyes are unique. As a consequence, your eyes are all distinct.

The iris is detected using a video camera before being decoded and stored. Once the data is stored in the computer system, retrieving it in milliseconds is easy when a person looks into the camera.

Iris identification offers a variety of advantages over other biometric identification methods. To begin with, this technique of identifying an individual is novel. The features of your iris will never change regardless of your age. On the other hand, if you engage in intense physical labour, your fingerprints may change with time, making scanning them more challenging.

Indeed, these security solutions are AI-based. Face recognition technology may help eliminate the requirement for conventional fingerprint identification. As a consequence, this system is very user-friendly and hygienic.

Apart from that, it is much more accurate and faster than other biometric systems. In a couple of seconds, these devices will scan and verify your face. Consequently, workers will no longer be needed to utilize their cards, keys, or badges. Once deployed, these facial recognition systems can not be misused. These techniques will be very helpful if you are an employer.

A Facial Recognition System's Characteristics

Now, let's look at some of the most distinguishing features of a face recognition system.

• Prevents tampering with time • Extremely efficient

• Managing contract workers or temporary employees • Forensic investigators will appreciate the face crop feature.

IFF

• Multiple facial recognition • Increased reliability • And many more

In conclusion, this article provided an overview of face recognition systems for businesses of all sizes.

CHAPTER TWELVE

Educate Yourself about Patents before Filing a Patent Application

A patent is a government-issued document that bestows the holder with the exclusive right and permission to use or distribute an invention. Without the patent holder's authorization, the invention can not be commercially distributed for a set time.

Before submitting a patent application, it is critical to have a working knowledge of patents and their benefits.

Numerous patents

Patents are filed for a variety of inventions. Being familiar with the different kinds of patents will aid in obtaining the required rights and protection for your invention.

• Design: This is filed when an investor wishes to protect the shape or design of their creation. The document is illustrated and photographed extensively to show the beneficial invention's design.

• Utility: This is the technical word for a patent. It is a document that describes how a freshly developed machine or process works and operates. It contains essential information to educate the audience on how to use the invention.

• Plant: It protects any new plant that is not the result of sexual reproduction, such as cutting. This patent relates mostly to traditional horticulture. This patent excludes genetically modified species from its scope.

• Provisional: It provides an extra year to decide whether or not to file a utility patent application and how to proceed. It is a certificate that certifies the bearer created something and is halfway through figuring out how it works. This patent has a time constraint, requiring the holder to file for a utility patent within one year or risk forfeiting the filing date.

The advantages and disadvantages of patenting

If someone is undecided about applying for a patent, these advantages and disadvantages will help them decide.

Advantages

• The most beneficial feature is that the documentation legally prevents any other person from using the innovation. Without the authorization of the patent owners, the invention can not be produced, utilized, or sold commercially.

• Have the freedom to use inventions for personal gain.

• Assists in the establishment of a market free of competition.

• May produce revenue if the patent holder decides to sell the patent. Additionally, the patent holder may license the invention to allow for its use by others. Numerous companies depend solely on this revenue stream.

Disadvantages

• Because getting a patent is a long process, and it is possible that the invention's utility may dwindle or that another technology will be created to replace it.

• Annual payments must be paid on time to ensure the patent remains valid and does not expire. • The overall cost of getting a patent may be very high. As a result, one must guarantee that future earnings surpass the cost of the initial patent. When a patent application is submitted, some details about the invention become public, raising the likelihood of competitors.

Patent the SLAM solutions you've deployed and reap the benefits of a patented invention. Additionally, we provide services for acquiring VLAM and QSLAM patents. It is better to get the patent immediately before another innovator exploits the sole proprietorship's advantage.

CHAPTER THIRTEEN

Digital transformation is a process driven by people, not technology

According to The Economist, the most visible effect of the ongoing Covid-19 epidemic will be "the injection of data-enabled services into an increasing number of spheres of life." Businesses are expected to prioritize digital transformation soon.

A 2019 study of CEOs, directors, and certain senior executives found that their primary concern was the risk associated with digital transformation. However, 70% of their efforts to move this motion forward fell short of their goals. Regrettably, $900 billion of 2019's $1.3 trillion investment in new ventures was squandered.

Why?

Fundamentally, digital transformation teams fail despite the opportunities for growth and efficiency gains due to a lack of change-oriented individuals. It is extremely difficult to transform an organization with flawed organizational processes completely. Additionally, digitalization magnifies flaws, making them appear larger.

What is Digital Transformation?

When a business introduces a new system, it's natural to implement, specify, and count plans to get a little out of hand.

Digital transformation is critical to a business's continued relevance and profitability in today's competitive market.

The process entails incorporating new technology and services into existing business processes to streamline operations. The goal is to improve and add value to the finished product. It requires the addition of new tools and applications, data storage, information recording, and the adoption of numerous new methods.

Naturally, this is the digital aspect of the equation. However, if you consider it, we discuss "transformation," which introduces new ways of working with the current team.

That is a very deceptive statement!

While anyone would be willing to invest in a new set of digital suites equipped with cutting-edge technologies, who would manage them? Here, it is critical to ensure that talent, or people, is on board and that the business culture is adaptable. A successful transformation necessitates change management, which can only be accomplished through the efforts of individuals.

Include Your Colleagues

Any change is difficult. If you want to make significant changes in your organization, you must ensure that everyone is on board, not just your leadership team. While you can not allow your team to make significant decisions on your behalf, involving them in the process may improve outcomes.

Although 84 percent of CEOs are committed to significant transformation initiatives, only about 45 percent of frontline workers concur. Connecting the dots is a significant impediment to successfully implementing a plan.

There are many ways to do this:
- Collect feedback from your team regarding the changes you made.
- Inform your staff about the implementation plan.
- Employ internal marketing to convince even the most obstinate team members to adopt new technologies.

The transformation of an organization's digital environment may be beneficial, but only if all team members agree and embrace the change. Ascertain that your digital transformation team is optimistic and understands why embracing new technology and its benefits are critical.

Invest in the development of your team.

Going digital would present difficulties. Certain members of your team may be less technologically savvy than others. However, you can not abandon them. To assist them in adapting to new technologies and tools, extensive training is required.

Bear in mind that everyone learns in their unique way and at their own pace. For instance, some team members may grasp the concept after a single demo session, while others may require several days of training to become proficient with the new technology.

Experiment with various training resources, such as online courses and hands-on learning, and provide learners with various learning options.

It may take time to understand how to use new technology to achieve better results, especially for team members who lack an innate affinity for technology. Investing in training enables you to expedite this process.

The Digital Transformation Framework Does Not Change Everything.

The fundamental tenet of digital transformation is not to revolutionize everything simultaneously.

When you begin to transform your business, it's easy to get carried away. It is critical, however, to understand the technologies that will be implemented. Consider which method is most likely to be adopted by workers and select the optimal method.

Anything that glistens is not always preferable. When you choose to transform your business's operations digitally, it is solely to streamline work processes and make life easier for your team members. As a result, avoid making things complicated. Consult the frontline staff if you have any questions about the changes.

For example, consult your staff if you want to implement a new platform for online communication but are undecided between Zoom, Teams, and Slack.

Broaden Your Aspirations

When it comes to major transformations, avoid myopia. The goal of digital transformation services is to simplify and improve people's lives. A successful transformation strategy entails introducing new processes and procedures to increase efficiency and decrease employee workloads.

If properly executed, such a digital revolution could result in improved work practices, increased value for customers, and decreased workload for the team. If your digital move does not meet all of these criteria, something is wrong.

Change Must Begin at the Top

Grassroots change is a self-evident concept. In reality, change is more likely to occur if it is initiated from the top. Again, this does not imply a hierarchical or autocratic structure or a fear-inducing culture. It simply implies transformational and transactional leadership.

When it comes to digital change, the primary implication is that no significant change or even an upgrade to the organization is possible unless and until top leaders are selected and developed. It is self-evident that good and bad leadership has a cascading effect on every facet of an organization. The CEO or top leader of an organization is the single most important factor determining the effectiveness of its transformation. Naturally, industry, culture, context, legacy, people, and authentic technology are all critical, as are other resources.

However, these factors are quite similar across competitors, whereas senior-most leaders' values, mindset, integrity, and competence serve as the primary differentiator. Naturally, while everything else in an organization can be replicated, talent can not. Therefore, invest in the best talent to maximize impact, where the greatest value is generated.

Finally, a few words

While technology is all about achieving more with fewer resources, this strategy works best when combined with human capabilities. As was the case with technological disruption, which resulted in automation and the abolition of obsolete jobs, resulted in new ones. That is why innovation is also known as creative destruction. ' Any creative aspect of innovation is contingent upon people. Thus, technology and humans can coexist by leveraging human adaptability to upskill and reskill the workforce. Simply put, a brilliant innovation is meaningless if there is low-skilled labour to implement it, and even the most inspiring human minds are ineffective if they are disconnected from technology. The major implication is that when leaders consider new technology, they should also consider the people who make it useful.

CHAPTER FOURTEEN

AngularJS's Complete History

It was founded over a decade ago and has seen many changes since then. The first version of the framework was discovered in 2009, shortly after the launch of AngularJS, and therefore laid the groundwork for modern application development.

Misko Hevery, a programmer, founded AngularJS after taking on a side job. Misko constructed a framework to address the shortcomings of all HTML while shooting ideas and best practices from many libraries that had previously been used to do various tasks. Misko's approach to legacy Angular was a brilliant execution due to the characteristics that made it very popular with additional web professionals.

AngularJS is the name given to the 1.x and 2.x versions of this frame. Angular was not only the framework in its early days; it also had a lot of the most helpful methods and features, which quickly made it popular. After gaining popularity alongside other frameworks in a matter of months, it drew the attention of Google, which saw the tremendous potential of the Angular framework created by its team. Thus, Angular gained popularity as a result of the support of massive businesses such as Google.

Before the arrival of Angular on the scene or during its early days, it was not simple to manage the large bundle size compared to other libraries. There were many performance issues in the frame solved by several people. Numerous flaws were discovered in the frame, reducing the capability of AngularJS. The Angular frame's template syntax performed well, and Vue.js adopted it. (v-if-ng-if, Vmodel-ng-model) Structure, whereas the chaotic eat-up loop mechanism imposes limitations on its speed.

Some delays in AngularJS served as an excellent motivator, compelling it to reveal the complete frame. Because Google's programmers desired to reveal the frame, developers began using very strong libraries. Though AoT converts HTML (hypertext Markup Language) and TypeScript code to JavaScript during development, tree-shaking removes unnecessary imports to enable quicker application boot-up and a lower footprint.

Indeed, the CLI debuted to start new projects, create skeletons, and build application servers, which has resulted in it being a very useful tool for programmers. Apart from that, Angular has several helpful mechanisms for managing browser history. As a consequence of these mechanisms, it is now possible to deal with URL changes through direct client interaction and perhaps the browser's back/forward buttons. When an Angular program is an input, an abstract path is created, after which URLRouterProvider may be described as the default route. Through the HTML 5 history API and Angular's location support, we may even gain control of the browser.

Almost six decades ago, the Angular team acquired a plug-in for its Google-Chrome browser called Batarang, primarily designed for online software debugging. The expansion is incompatible with Angular's post-release variations.

CHAPTER FIFTEEN

The Prince2 Agile Foundation Certification Is Valuable

PRINCE2 Agile has evolved into the world's finest and most complete project management system, combining the adaptability and flexibility of agile with the well-defined framework of PRINCE2. The unique training guideline shows how institutions employing agile and PRINCE2 may use this compatibility, arming them with the skills and capabilities necessary to complete projects that address consumer demands in rapidly changing work environments.

- Achieves a balance of strength and adaptability
- Detailed descriptions of projects
- Seamless integration
- Increased ability to respond and adjust

Overview

To get the PRINCE2 Agile Foundation certification, candidates must complete a multiple-choice exam.

However, applicants must first attend a three-day mentorship program led by industry experts. Anyone may begin from scratch if they have no prior experience. Participants may get a thorough understanding of agile workgroups, administration, and execution. To get the certificate, candidates must demonstrate their understanding of each of these concepts correctly by answering the exam questions.

Prospective candidates

PRINCE2 Agile seems to be a good fit for anybody managing large or small projects in an agile setting. It is a valuable certification and qualification program for anybody working in an agile project environment, whether as a project coordinator, support service provider, or part of a larger project group. The PRINCE2 Agile Foundation requires few preconditions. Those with no previous knowledge of PRINCE2 may sit for the test since the methodology is taught during the course. The PRINCE2 Agile Foundation certification qualifies professionals to manage agile projects using PRINCE2 monitoring systems. Such laws rely on agile supply chain methods, which need both a broad toolset and a framework. It allows companies to have more control over the flexibility of projects, thus increasing income.

Techniques and Suggestions for Exam Preparation

• **Enroll in a PRINCE2 training course.**

It may help you earn the required PDUs for the PRINCE2 Certificate Program. Additionally, a candidate will get a PRINCE2 certificate with progressive effectiveness by integrating optimal materials and requirements. However, before starting professional education, study the full PRINCE2 manual.

• **Provide oneself with an ideal learning environment.**

It's challenging to study while working as a professional. As a consequence, the applicant must maintain concentration and thorough preparation. You must spend the maximum amount of time possible studying while avoiding distractions.

• **Create and adhere to a study schedule.**

To begin, develop a detailed study schedule that covers the majority of the curriculum's topics. Along with learning, take notes and ask questions. Prefer to maintain a schedule as much as possible, but recognize that too much of anything is detrimental.

• **At all times, be informed and connected to a PRINCE2 network.**

Candidates may communicate with other PRINCE2 exam candidates via readily available online chat rooms on LinkedIn, Facebook, or Twitter. It may help you remain informed about current events and PRINCE2 Certification changes. Additionally, candidates may review the most current PRINCE2 publications written by certified experts.

• **Investigate PRINCE2 simulations**

Using simulations may help educate the mind to process information rapidly, which is particularly beneficial under exam stress. Establish a habit of gathering important points throughout each session to understand better the subjects on which applicants want to advance.

• **Consider PRINCE2 certification as a job.**

Finally, candidates are advised to approach their PRINCE2 Test preparations as a task. Simply said, set a goal and stick to it. It may give applicants more time to practice each subject covered in their study plan.

Are you interested in learning more about the benefits of project management for your business? Our experienced team of experts can evaluate your business's health. The moment has come to contact our Advisory team to prepare for the PRINCE2 Agile Foundation Exam.

CHAPTER SIXTEEN

Obtain Linksys Router Support

Linksys support is a well-known name in the field of online technical assistance. Due to their extremely useful function, routers have quickly become one of the most useful equipment. A wi-fi router essentially provides a wireless internet connection within a specified radius, such as your house or business. Their previous avatars, dubbed strained routers, were responsible for providing network access to a few people through an internet connection. However, stressed-out routers were harmed by significant disadvantages that were eliminated by wi-fi routers. The main drawback was that consumers needed to connect through wires to the router to access the internet. The second one was that, since a limited number of ports were provided, only a limited number of people could simultaneously access the internet in a single pass.

The significance of Linksys router support

Linksys routers do sometimes cause issues for their customers, prompting them to seek assistance. One of the most common issues that router users have is setting up the router and synchronizing it with their current internet connection. The second most prevalent issue that necessitated Linksys router assistance is related to motive force problems. Routers of different types need a unique collection of drivers, which should be updated regularly. Apart from them, many problems may result in the customer needing router assistance.

Obtain Router Support Immediately

Regardless of whether it's a Linksys router setup issue, a Linksys motive force issue, or another issue, you may enroll in the services of online technical help groups and enjoy independence from all your router problems. These businesses provide various apps that enable you to care for your wants while preserving your valuable wants. To subscribe, visit their website or call their toll-free lines to speak with a representative about the best available plan.

PC Care's expansive service portfolio includes Linksys router support, operating system & software program assistance, email & browser assistance, and assistance with the installation of all peripheral devices on your computer.

Our goal is to keep you comfortable and your computer is running smoothly by providing you with customized online computer support services that address the overall health of your laptop.

CHAPTER SEVENTEEN

Crystal Reports Invoice Delivery: How to Handle It?

"Whatever does not kill you strengthens you."

Over the past several decades, software engineers have taken this mantra to heart and strived for perfection. When discussing the Crystal Report, there is an air of caution around its use. Not because it made creating and producing reports more difficult, but because they wanted it to have more features, a nicer interface, less clutter, no archive, and more complete widgets.

One may either lament the inadequacies or begin brainstorming new, feasible solutions to the issues. These demerit points may ultimately be reduced with the use of a Crystal Reports busting tool. Without a doubt, the capacity of this automated tool will determine the total functioning of the program; thus, one should look for appropriate characteristics. However, this is an unsurprising situation to be in.

Systems for Automated Document Delivery

Is there any possibility of a software tool? Indeed, it is. Perhaps after learning some codes and fixes, you can create your solution. However, it occurs later, most likely after research.

What you need to establish is the following.

Apart from routine activities, you'll require a tool that can handle bulk functionality. The reports you produce will be required to be delivered to each recipient on time and without mistake. While Crystal Report Writer will certainly assist you in producing visually appealing reports, you may not discover a better function for sending them unless you have a flexible software application.

Invoice generation, although critical, should not stifle your workers' ability to accomplish much more. Small companies have many challenges regarding staff management, company operations, client management, and customer service, to name a few. Without the automatic invoice distribution system provided by Crystal Reports, many thriving companies would have joined the problematic lot.

Numerous forums have addressed the critical importance of storing Crystal Report functions for future usage. That is why you should look for solutions that have archiving capabilities. It may take some time to locate instruments that meet these requirements, but the effort is well worth it.

Methods that are simple to use and simulate for everyone. Indeed, the tool should not require you to master coding or other advanced techniques. It may be counter-productive for workers looking for a simple automated solution to handle daily paystub or invoice distribution chores. A trial version of the instrument would be very helpful for evaluating its performance and efficiency over a short period.

You need certain advantages.

Remember that you are purchasing a product designed to make tasks easier, faster, and more efficient. If the tool is anything other than this, you should go on to another. Numerous owners remain loyal to their current tools rather than experimenting with new ones available in the software market.

With a flexible document bursting tool, you may create and suppress document distribution according to your requirements. As a result, you'll save time, money, and arduous physical work.

Jesse Chris has worked with various Crystal Reports bursting tools and has seen various Paystub Delivery software systems. Writing about the document delivery techniques accessible today, he believes that companies have reaped several advantages due to this instrument. He continues by stating that the Crystal Report invoice delivery technology has simplified life for end-users.

CHAPTER EIGHTEEN

IOS Application Development Can Help You Expand Your Business Prospects

The digital technology race is on; recent advancements in digital technology, especially smartphones, have transformed. Smartphones are the most effective instrument for increasing your business's sales. Businesses create flexible mobile apps to engage their audience. However, people often do not know which operating system to select since each OS has unique functions and characteristics.

Since the dawn of the age of flexible applications, consumers have been swayed by two primary forces: iOS and Android. iOS is Apple's mobile operating system, which powers the iPad, iTunes, iPod touch, and the newly released Apple TV. iOS offers a plethora of adaptable platforms that provide a unique environment for developing bespoke business applications. It has an active marketing approach, which is seen in its iOS apps. iOS apps are designed just for them to collect and organize data to make critical business decisions. Businesses may benefit from iOS apps by providing critical information and data to their representatives whenever and wherever they need it. Additionally, an iOS application may help a company increase its sales and profits.

The primary motivations for developing business applications for iOS are as follows:

Secure Access:

Apple maintains a tight grip over the whole ecosystem, from hardware to firmware to programming. That means the company thoroughly checks each program that appears in its application store, significantly reducing the danger of downloading malicious or surreptitious apps. iOS devices offer excellent legacy support, ensuring that your device always runs the most current software with the most up-to-date security patches.

Versatile uses of superior quality:

The primary reason for iOS development's enormous success is because the platform is incredibly focused on the client experience. It offers unmatched capabilities and functionality to iOS users.

Deployment and management are simplified:

iOS applications are very simple to deploy and manage on a wide scale, allowing each iPhone, iPad, and Mac to be set up and configured automatically. It obviates the need for an IT staff to administer and maintain it. Additionally, IT may send apps directly to devices, and salespeople can modify company-owned Apple devices by adding their applications.

Compatibility with your platform:

iOS is more compatible than other operating systems, enabling iPhone application developers to focus on the program's development rather than application compatibility problems. Apple devices offer advanced capabilities, a faster CPU, and enough storage space, enabling iOS application developers to build sophisticated apps that adhere to user interface standards.

We've covered how iOS app development may help your company grow and the primary arguments for choosing iOS to build your business applications. To empower your business in the global market, you must select the best development company, as a professionally developed iOS application enables you to endorse your brand or business, increasing your market availability, protecting your customer information, engaging your clients with your services and products, and increasing productivity through viability services to reach your targeted users.

CHAPTER NINETEEN

Can the Crestron Home Automation System be integrated with Sonos?

Whether you're contemplating installing a Crestron House Automation system in your house or already have one, you may be wondering whether connecting Crestron to Sonos is possible. Crestron is, without a doubt, the industry leader in home automation, and many customers would want to connect their system with Sonos.

Crestron and Sonos systems are compatible. Both companies recently collaborated to ensure the successful integration of the two systems. Custom Controls, a well-known home automation installation company in the United Kingdom, confirmed that the two systems function well together.

They said that it is entirely possible to launch the Sonos app straight from inside Crestron, avoiding the need to switch apps. It increases the usability of collaborative rather than competitive technology. Additionally, you may link Sonos to an existing Crestron panel through a Sire gateway.

The Sonos ecosystem is a great way to enjoy music in crystal clear surround sound across your home. Crestron's audio transmission enhances this experience even further.

Additionally, Sonnex technology is available, which is a high-end premium audio distribution system created by Crestron. Many of you will have never heard of this method. It mixes several audio channels and calibrates rooms, with the possibility of adding subwoofers to individual audio zones for a truly modern home music system.

Additionally, if required, the system may be easily scaled up. For instance, Custom Controls recently completed a project in the United Arab Emirates. They installed an astonishing 72 zones of music, each zone capable of enjoying a varied variety of audio sources through Sonos.

We think that the Sonos system's flexibility and growing popularity, coupled with Crestron's cutting-edge home automation technologies, work extremely well to bring your house into the modern age. We highly encourage you to have both systems installed in your home.

Have you previously incorporated this feature into your home? If this is the case, we'd love to hear from you.

Bibliography

Bibliography

Shalini M. (2021, July 15). The Importance of Data Science As a Career. EzineArticles; EzineArticles. https://ezinearticles.com/?The-Importance-of-Data-Science-As-a-Career&id=10474966

Shalini M. (2021, July 19). Understanding Artificial Intelligence - Career, Admissions and Requirement in Australia. EzineArticles; EzineArticles. https://ezinearticles.com/?Understanding-Artificial-Intelligence---Career,-Admissions-and-Requirement-in-Australia&id=10474969

Shalini M. (2021, July 19). Factors To Consider Before You Buy Cable Glands. EzineArticles; EzineArticles. https://ezinearticles.com/?Factors-To-Consider-Before-You-Buy-Cable-Glands&id=10487532

Shalini M. (2021, July 19). Tips To Help You Choose The Size Of Cable Glands For Your Project. EzineArticles; EzineArticles. https://ezinearticles.com/?Tips-To-Help-You-Choose-The-Size-Of-Cable-Glands-For-Your-Project&id=10487538

Shalini M. (2021, July 19). An Introduction to a Cable Gland. EzineArticles; EzineArticles. https://ezinearticles.com/?An-Introduction-to-a-Cable-Gland&id=10487541

Shalini M. (2021, July 19). Why Do You Need Cable Glands For Your Electrical Equipment? EzineArticles; EzineArticles. https://ezinearticles.com/?Why-Do-You-Need-Cable-Glands-For-Your-Electrical-Equipment?&id=10487550

Monroe, M. B. (2021, July 23). Why Did Banks Ban Cryptocurrency Purchases Using Their Credit Cards? EzineArticles; EzineArticles. https://ezinearticles.com/?Why-Did-Banks-Ban-Cryptocurrency-Purchases-Using-Their-Credit-Cards?&id=9987619

6 Features of a Good 4K Portable Monitor. (n.d.). EzineArticles. Retrieved August 25, 2021, from https://ezinearticles.com/?6-Features-of-a-Good-4K-Portable-Monitor&id=10489944

What Is the Future of Artificial Intelligence Courses in South Africa? (n.d.). EzineArticles. Retrieved August 25, 2021, from https://ezinearticles.com/?What-Is-the-Future-of-Artificial-Intelligence-Courses-in-South-Africa?&id=10474975

Shalini M. (2021, July 27). The Introduction Of The Face Recognition System Saudi Arabia. EzineArticles; EzineArticles. https://ezinearticles.com/?The-Introduction-Of-The-Face-Recognition-System-Saudi-Arabia%E2%80%A8&id=10490010

Shalini M. (2021, July 28). Know More About Patents Before Applying For One. EzineArticles; EzineArticles. https://ezinearticles.com/?Know-More-About-Patents-Before-Applying-For-One&id=10491171

Saurav Mittal. (2021, August 4). Full History on AngularJS. EzineArticles; EzineArticles. https://ezinearticles.com/?Full-History-on-AngularJS&id=9988099

Shalini M. (2021, August 9). Value of Prince2 Agile Foundation Certification. EzineArticles; EzineArticles. https://ezinearticles.com/?Value-of-Prince2-Agile-Foundation-Certification&id=10496088

Johnson, A. (2021, August 9). How To Get Linksys Router Support? EzineArticles; EzineArticles. https://ezinearticles.com/?How-To-Get-Linksys-Router-Support?&id=9988336

Chris, J. (2021, August 19). How To Deal With Crystal Reports Invoice Delivery. EzineArticles; EzineArticles. https://ezinearticles.com/?How-To-Deal-With-Crystal-Reports-Invoice-Delivery&id=9988411

Rism, C. (2021, August 19). Miami Web Design Company and SEO Expert Services in the Doral Area. EzineArticles; EzineArticles. https://ezinearticles.com/?Miami-Web-Design-Company-and-SEO-Expert-Services-in-the-Doral-Area&id=9988522

Amenda Cerny. (2021, August 19). Boost Your Business Prospects With IOS App Development. EzineArticles; EzineArticles. https://ezinearticles.com/?Boost-Your-Business-Prospects-With-IOS-App-Development&id=9989206

Jackson, K. (2021, August 20). Can the Crestron Home Automation System Be Integrated With Sonos? EzineArticles; EzineArticles. https://ezinearticles.com/?Can-the-Crestron-Home-Automation-System-Be-Integrated-With-Sonos?&id=9989473